ULTIMATE SPECIAL FORCES
US NAVY SEALS

TIM COOKE

FARGO PUBLIC LIBRARY

PowerKiDS press

New York

Published in 2013 by The Rosen Publishing Group, Inc.
29 East 21st Street, New York, NY 10010

Copyright © 2013 Brown Bear Books Ltd

All rights reserved. No part of this book may be reproduced in any form without permission in writing from the publisher, except by a reviewer.

Senior Editor: Tim Cooke
US Editor: Sara Antill
Designer: Supriya Sahai
Picture Research: Andrew Webb
Picture Manager: Sophie Mortimer
Creative Director: Jeni Child
Children's Publisher: Anne O'Daly
Production Director: Alastair Gourlay
Editorial Director: Lindsey Lowe

PICTURE CREDITS
Front Cover: US Navy
Corbis: Peter Souza 37; Xinhua Press 36; Getty Images: AFP/Roberto Schmidt 35tl; Library of Congress: 24, 25; Robert Hunt Library: 04, 05, 27, 45t; US Army: 31; US Department of Defense: 06, 07, 33bl, 34; US Marine Corps: 40/41; US Navy: 08, 09, 10, 11, 12/13, 13tr, 14, 15, 16, 17tr, 17br, 18, 19, 20, 21, 22, 23, 26, 28, 29, 30, 32, 33tr, 35b, 38, 39tl, 39br, 41br, 42, 43, 44, 45bl.
Key: t = top, c = center, b = bottom, l = left, r = right.

Library of Congress Cataloging-in-Publication Data

Cooke, Tim, 1961–
 US Navy SEALs / by Tim Cooke.
    p. cm. — (Ultimate special forces)
 Includes index.
 ISBN 978-1-4488-7881-9 (library binding) — ISBN 978-1-4488-7958-8 (pbk.) —
ISBN 978-1-4488-7963-2 (6-pack)
 1. United States. Navy. SEALs—Juvenile literature. 2. United States. Navy—Commando troops—Juvenile literature. I. Title.
 VG87.C66 2013
 359.9'84—dc23
                              2012011752

Manufactured in the United States of America

CPSIA Compliance Information: Batch #B2S12PK: For further information, contact Rosen Publishing, New York, New York, at 1-800-237-9932.

CONTENTS

INTRODUCTION 4
WATER WARRIORS 6
ORGANIZATION 8
DELIVERY VEHICLE TEAM 10
SPECIAL BOAT TEAM 12
SELECTION 14
BUD/S TRAINING 16
UNDERWATER TRAINING 18
SEAL TEAM SIX 20
INSERTION AND ESCAPE 22
VIETNAM 24
PANAMA 26
AFGHANISTAN 28
OPERATION RED WING 30
IRAQ 32
SOMALIA 34
PAKISTAN 36
BOATS 38
AIRCRAFT 40
SPECIALIST GEAR 42
WEAPONS 44
GLOSSARY 46
FURTHER READING AND WEBSITES 47
INDEX 48

Introduction

SEALs is a good acronym for the US Navy special forces. They began in the water and are still at home there. But SEALs stands for Sea, Air, and Land teams and modern SEALs are trained to fight anywhere. They're one of the most secret fighting forces. But in 2011 SEAL Team Six achieved international fame when it killed the al-Qaeda terrorist leader Osama bin Laden in Pakistan in a night raid.

AN EARLY combat swimmer prepares to test his breathing apparatus.

INTRODUCTION

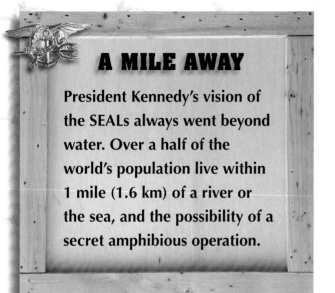

A MILE AWAY

President Kennedy's vision of the SEALs always went beyond water. Over a half of the world's population live within 1 mile (1.6 km) of a river or the sea, and the possibility of a secret amphibious operation.

GROWING ROLE

The raid in Pakistan was in many ways just the sort of task for which the SEALs were created, but it took place far from the ocean. The force started by President John F. Kennedy in 1962 was a small elite team meant to carry out unconventional maritime warfare. Since then it has grown to include antiterrorism roles and other activities. Today SEALs can be found in the deserts of Afghanistan or on the streets of rebel-held cities, as well as on or in the water.

A MEMBER of a Navy Demolition Unit prepares to blow up beach obstacles in World War II.

US NAVY SEALS

Water Warriors

The roots of the SEALs lay in World War II (1939–1945). In Africa, Europe, and the Pacific, US troops needed to make amphibious landings in enemy-held teritory. The landings faced obstacles such as tank barriers and mines.

WEARING FROG suits, combat swimmers train for landing on enemy beaches in an inflatable boat.

COMBAT DEMOLITION UNITS

The US Navy trained a new force to go ashore secretly, find out about the enemy, and locate and destroy obstacles. These combat swimmers formed first the Naval Combat Demolition Units, commissioned in October 1942. The following month they took part in Operation Torch, the first Allied landings in occupied North Africa. Also known as Underwater Demolition Teams (UDTs), they worked in groups of six men.

ON DRY LAND

Since 9/11, SEALs have increasingly operated far from water, in the deserts of Iraq and the mountains of Afghanistan. Their small, highly trained units are better for antiterrorist activities than conventional forces.

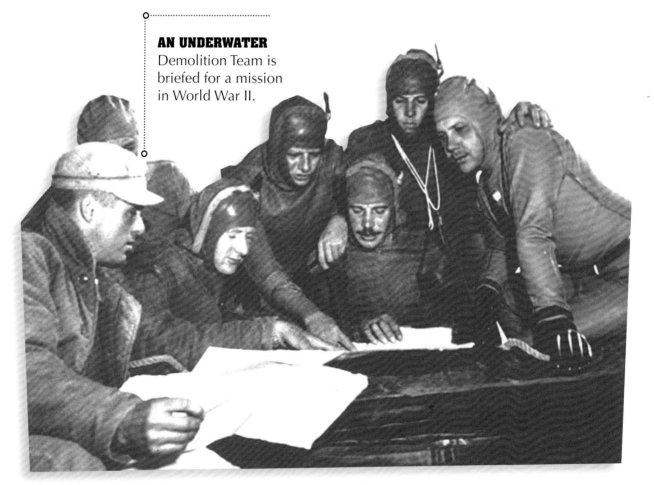

AN UNDERWATER Demolition Team is briefed for a mission in World War II.

Organization

The SEALs punch above their weight. They're only a small part of the US Navy and they total fewer than 2,300 men. But from their bases—Little Creek, Virginia, on the East Coast and Coronada, California, on the West Coast—they can undertake missions virtually anywhere.

FIVE ROLES

SEALs are trained for five basic types of mission:
- Unconventional warfare
- Foreign internal defense
- Direct action
- Counterterrorism
- Special reconnaissance

A SEAL comes ashore. SEALs are trained to fight as soon as they hit the beach.

SECRET STRUCTURE

Much information about the SEALs is classified, or secret. We're not even sure how many SEAL teams there are. There are at least eight, but there may be more. Each team has a total of about 130 men, including its support staff. Each team has six platoons of sixteen men: two officers, one chief, and thirteen enlisted men. A platoon usually operates in smaller units of four or five men. Every member of the platoon can do a range of tasks. All SEALs have to be qualified in diving, parachuting, and demolitions.

SEALS practice techniques for boarding oil and gas platforms.

US NAVY SEALS

DELIVERY VEHICLE TEAM

The SEAL Delivery Vehicle (SDV) Team is a specialized unit that gets other SEALs to wherever they need to be to carry out a mission. They operate the MkVIII SDV minisub. The SDV Team can be exposed to water for long periods and have to use scuba equipment.

SDV TEAM members inside a shelter mounted on the deck of a submarine. Such shelters allow subs to deploy SEALs without surfacing.

DELIVERY VEHICLE TEAM

SDV TEAM divers practice lock-out maneuvers for exiting a submarine deep underwater.

EYEWITNESS

"The forward torpedo room was designed to accommodate six men, who loaded the torpedoes into the forward tubes. We filled the space and then some. Sweat filled my wet suit and dripped off my face."

Bob Gormly
SEAL captain, UDT,
Vietnam, 1966

GETTING TO THE ACTION

SDV Teams get SEALs as close as possible to the location of their mission without being detected. Each two-man crew can carry four passengers and their equipment. They take SEAL teams from submarines or warships to within 2 miles (3.2 km) of their destination. From there, the SEALs can swim if they have to. The SDVs are vital. Without them, SEALs would be limited to working only within the immediate areas around subs or ships.

Special Boat Team

Special Boat Teams (SBTs) carry out patrols close to shorelines, as well as along rivers, in high-performance boats and small ships. They're not SEALs, but they work closely together. SBTs deliver SEAL teams for their missions and get them out again at the end.

A SPECIAL Boat Team practices picking up SEALs from river banks.

JOINING SBT

Special Boat Teams are part of the Naval Special Warfare Group, like the SEALs. Recruits have to be in great shape. It takes a 22-week training course to qualify as a SWCC (Special Warfare Combatant-Craft Crewmember).

SPECIAL BOAT TEAM

SPECIAL Boat Team 12 trains offshore. These fast boats are armed with machine guns.

BROWN WATER NAVY

The idea for the SBTs came during the Vietnam War (1961–1975). Navy commanders realized they needed patrols off the coasts and on Vietnam's many rivers, what they called a "Brown Water Navy." With fast boats that operated in shallow water, the SBTs helped block the North Vietnamese from delivering supplies. Today the three SBTs have a total of about 700 sailors. They are trained in handling fast boats, first aid, and ocean survival.

13

SELECTION

Any male in the US Navy age 28 or younger can apply to be a SEAL, but only if he passes the PST, or physical screening test. A SEAL must be able to swim 500 yards (460 m) of breast or side stroke in 12.3 minutes, run 1.5 miles (2.4 km) in 11 minutes, and do six pull-ups, 42 push-ups, and 50 sit-ups. And then he must do it all faster just a few weeks later!

A TEAM OF recruits perform surf passage training with a small inflatable boat.

SEAL PREP COURSE

Candidates spend eight weeks in Chicago, Illinois, on the SEAL Prep Course. This first round of training is an introduction to the even tougher Basic Underwater Demolition School (BUD/S) that comes next. At the end of the course, candidates take another PST. If they have not improved on their original scores, they fail. Throughout the whole process, the drop-out rate is high. Only the candidates in the best mental and physical shape can cope with the demands of being SEALs.

SELECTION

- 5–9-week prep course
- 3-week BUD/S orientation
- 7-week BUD/S Phase I physical conditioning training
- 7-week BUDS/S Phase II combat diving
- 7-week BUD/S Phase III land warfare
- 3-week parachute jump school
- 26-week Seal Qualification Training (SQT)

A CANDIDATE'S swimming skills are tested in the swimming pool and in the ocean.

US NAVY SEALS

BUD/S TRAINING

Basic Underwater Demolition School (BUD/S) is the 26-week selection process to become a Seal. It tests candidates to the limit. About 60 percent drop out. In some classes, no one qualifies. Most drop-outs come in the last part of the first phase of training, often called Hell Week.

BUD/S candidates support a log to test physical strength and endurance.

BUD/S TRAINING

SURF PASSAGE, or getting boats or swimmers through the surf, is a physically demanding part of the BUD/S program.

"The belief that BUD/S is about physical strength is a common misconception. Actually, it's 90 percent mental and 10 percent physical. [Recruits] just decide that they are too cold, too sandy, too sore, or too wet to go on. It's their minds that give up on them, not their bodies."

BUD/S Instructor
San Diego

HELL WEEK

In Hell Week candidates train nonstop for five and a half days. They have just four hours of sleep in that whole time. They're always on the move; they're wet and cold, covered in mud or sand. They can drop out at any time by ringing a bell. But Hell Week teaches them to push their bodies to the limits and tests how they operate when they are exhausted.

CANDIDATES catch their breath during a swimmer surf passage exercise.

Underwater Training

The first SEALs were frogmen, or combat swimmers. Today, many missions are still carried out underwater. SEALs practice wearing scuba gear for days at a time. More important, they learn to stay underwater for long periods without equipment. To do this, they have to train themselves to hold their breath.

A CANDIDATE for a Special Boat Team ties knots during underwater proficiency training.

UNDERWATER TRAINING

A SEAL uses a compass for an exercise in underwater navigation.

INTENSIVE PROGRAM

SEALs study everything about operating underwater. They learn what happens to the body when it is submerged for long periods and the dangers of diving at depths below 100 feet (30 m), when a condition called nitrogen narcosis, or the bends, can kill a diver. They practice dealing with equipment failure. They learn how to swim for hours in dark, cold water and how to attach explosives to underwater targets by touch alone.

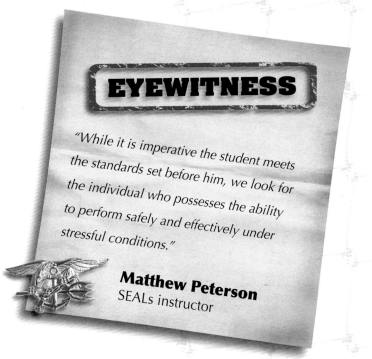

EYEWITNESS

"While it is imperative the student meets the standards set before him, we look for the individual who possesses the ability to perform safely and effectively under stressful conditions."

Matthew Peterson
SEALs instructor

US NAVY SEALS

SEAL Team Six

In 1980 an operation to rescue American hostages from the US Embassy in Tehran, Iran, ended in disaster. The helicopters carrying the rescue teams crashed. US military chiefs decided to create an elite force to carry out hostage rescues anywhere in the world. That job was given to SEAL Team Six, which was led by Richard Marcinko.

SEALS PROVIDE cover during an exercise in hostage rescue.

SEAL TEAM SIX

EYEWITNESS

"We were unique; a small, highly mobile, quick-reaction team trained to do one job: kill terrorists and rescue hostages, and do it better than anybody in the world. Nobody could move as fast as we could. No other unit could come out of the water or the sky with equal ease."

Richard Marcinko
Founder, SEAL Team Six

SECRET ELITE

SEALs who make it into Team Six train for fast in-and-out operations with helicopters. On the ground, they practice in situations with lots of close-range shooting. They have good language skills so they can go undercover. The identity of SEAL Team Six is a secret; so are most of its missions. In January 2012, however, it freed an American and a Dane from pirates in Somalia.

SEALS USE a mock-up building to train for close-quarter fighting.

Insertion and Escape

A NAVY SEAL makes a low-level jump from a Sea Hawk helicopter.

SEALs can get themselves anywhere for a mission and get themselves out again. They spend three weeks training as parachutists, who can jump over land or into water. They learn escape and evasion tactics. If they can't get away, they fight to the death. No SEAL has ever become a prisoner of war.

INSERTION AND ESCAPE

A SEAL freefalls high above the Arctic Circle in a winter warfare exercise.

EXTREME TRAINING

Other standard SEAL skills include mountain climbing, rappelling, and handling explosives. In extreme weather training, recruits go to Alaska. As part of the program, they break the ice to jump into freezing water without any wet or dry suit. They have to tread water for at least three minutes before they can get out and dry off.

SURPRISE

Surprise is key to the success of a mission. SEALs approach a coast underwater and creep ashore at night. Or they freefall from airplanes and open their parachutes at low altitudes. Or helicopters carry them close to the ground, too low to show up on enemy radar.

US NAVY SEALS

VIETNAM
1963–1975

The Vietnam War was the SEALs' first deployment in combat. Working in 14-man platoons, the SEALs went into Vietnamese cities held by the enemy. They ambushed enemy units, carried out reconnaissance, and rescued US prisoners. The SEALs proved highly effective, losing only 49 men in nine years of fighting.

MAP OF SOUTHEAST ASIA

A SEAL TEAM poses with its Vietnamese allies before a mission.

VIETNAM

MEKONG DELTA

SEAL Task Force 117 (the Mobile Riverine Force or MRF) fought in the Mekong Delta. This area of southern Vietnam had few roads but thousands of small rivers and canals. The MRF patrolled the waterways in heavily armed floating pontoons. Other SEAL teams went upriver to the jungle in speedboats to lay booby traps for the enemy.

THE "BROWN WATER Navy" in action: SEALs in a rigid-hull boat patrol a river in southern Vietnam.

EYEWITNESS

"Our ops at first were kind of touch and feel. We had a lot of fun. Us young lieutenants had tremendous freedom about how we wanted to run an op. We couldn't be told by anybody to run an op that we didn't want to do."

Bob Gormley
commander, SEAL Teams Two and Six, Vietnam

25

US NAVY SEALS

PANAMA
1989

In December 1989, Task Unit Whiskey, part of SEAL Team Four, checked their weapons one last time. They were about to begin a risky mission as part of Operation Just Cause, the US invasion of Panama. They had to blow up the presidential gunboat, *Presidente Porras*, to stop the dictator General Noriega from fleeing by water.

MAP OF PANAMA

MEMBERS OF SEAL Team Four pose just before taking part in Operation Just Cause in Panama.

PANAMA

A SOLDIER looks at the wreckage of Noriega's jet after the raid on the airfield.

AT THE AIRFIELD

While the gunboat was being sunk, a second SEAL unit headed to the airfield. There was a firefight with Panamanian guards. Four SEALs died. But the team successfully destroyed Noriega's personal jet. The dictator would have no escape.

SUCCESSFUL MISSION

The mission was a perfect example of a SEAL underwater task. It involved night swimming for long periods, undetected, to reach the target. The combat swimmers, staying underwater, had to avoid enemy fire, place explosives on the underside of the boat, and get away before they detonated. The mission could have gone wrong at any time. But it didn't and the SEALs achieved their objective. This was the first combat swimmer mission since World War II where the details were made public.

NAVY SEALS approach a village in Kandahar province at night to arrest insurgents.

AFGHANISTAN
2001-2002

SEALs were some of the first US forces in Afghanistan at the start of the War on Terror. Using desert patrol vehicles or traveling on foot, they carry out reconnaissance on the enemy Taliban. In January 2002, SEALs found and destroyed a huge Taliban arms dump in a network of caves in the mountains.

MAP OF AFGHANISTAN

AFGHANISTAN

QUICK RESPONSE

In February 2002 a US Predator drone spotted a Taliban leader, Khairullah Khairkhwa, leaving a building. Within 15 minutes, a team of SEALs and Danish commandos were taking off from Camp Rhino in a Pave Low helicopter accompanied by an Apache gunship. Less than 90 minutes after the first report, the mission was over and Khairkhwa was in US custody.

ON WATCH

SEALs spend weeks on patrol in the mountains of northeastern Afghanistan. The Taliban are active in the area, where they hide out in caves. The nearby border with Pakistan is a route for smuggling arms to the Taliban.

A SEAL watches as a Taliban arms dump is destroyed in the mountains of eastern Afghanistan in 2002.

Operation Red Wing

In 2005, men from SEAL Team Ten flew to Afghanistan. Their task was to locate Taliban and al-Qaeda terrorists responsible for an increase in casualties among US troops in the region. Operation Red Wing was intended to kill the terrorists and put a stop to the loss of US lives.

MEMBERS OF SEAL Team Ten pose before the operation. All but Marcus Luttrell (third from right) died in the firefight or the rescue mission.

COMBAT DISASTER

The mission went very wrong. The terrorists learned the SEALs' position. A gunfight broke out between four SEALs and more than 200 Taliban. Only one SEAL survived. To make things worse, a helicopter that went to the SEALs' rescue was shot down by the Taliban. Eight more members of SEAL Team Ten and eight Army special forces onboard died. It was the SEALs' worst ever day.

A SEAL stands sentry during an anti-insurgent operation in Afghanistan.

EYEWITNESS

"This was a border hot spot, where multiple Taliban troop movements were taking place on a weekly, or even daily, basis. We observed the Taliban way below us on the narrow, treacherous path through the mountains, moving along with their swaying camels loaded up with explosives, grenades, and God knows what else."

Marcus Luttrell
SEAL, Operation Red Wing

US NAVY SEALS

IRAQ
2003

MAP OF IRAQ

SEALs were already in the Persian Gulf before the US-led invasion of Iraq in 2003. With their inflatable boats, they had been clearing mines from rivers and harbors. But with the invasion on March 20, 2003, the SEALs took center stage in the first operation of the war.

EYEWITNESS

"They (the Iraqis) didn't know we were there until we were on top of them, and many of them were very happy for us to arrive."

Unidentified SEAL
Iraq, 2003

SEALS in a rigid hull inflatable boat sail alongside a US Navy cruiser.

PREVENTING DISASTER

US commanders worried that retreating Iraqi troops might set fire to oil facilities in southern Iraq and cause an environmental disaster. SEAL teams swam up to offshore oil platforms and seized them. Other SEALs helicoptered into the Al Faw oil refinery. Although surrounded by 3,000 Iraqis, the SEALs captured the refinery in a six-hour battle and held it until backup arrived.

SEALS form a visit, board, search, and seizure team to intercept suspect vessels.

AMONG OTHER SEAL targets was the refinery at Basra, damaged by the Iraqis.

US NAVY SEALS

Somalia
2009

In the early twenty-first century, Somali pirates from the east coast of Africa took many foreign ships hostage for ransom. In April 2009, pirates seized the US-flagged cargo vessel *Maersk Alabama* and took its captain, US citizen Richard Phillips, hostage. The kidnappers held Phillips in a lifeboat towed behind the cargo ship.

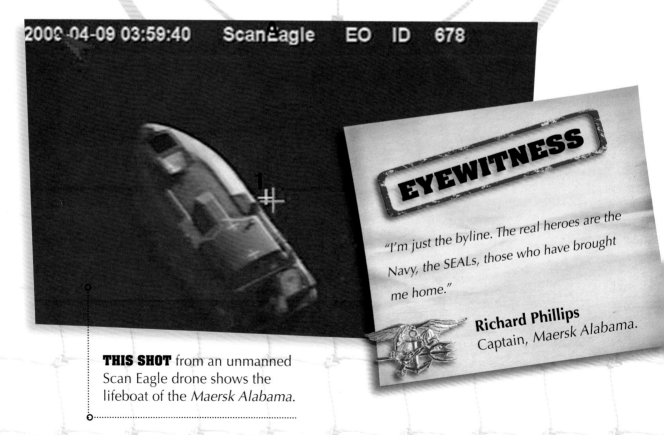

MAP OF SOMALIA

THIS SHOT from an unmanned Scan Eagle drone shows the lifeboat of the *Maersk Alabama*.

EYEWITNESS

"I'm just the byline. The real heroes are the Navy, the SEALs, those who have brought me home."

Richard Phillips
Captain, Maersk Alabama.

SNIPER KILL SHOTS

The US raced a SEAL team to the Indian Ocean. They parachuted into the sea then boarded the destroyer USS *Bainbridge*. The pirates agreed that the *Bainbridge* would tow the lifeboat while they negotiated the captain's release. That night, three SEAL snipers on the deck of the destroyer shot and killed the three pirates from 25 yards (23 m) away—a great feat from a moving ship. Captain Phillips was rescued

CREW MEMBERS of the *Maersk Alabama* celebrate news of their captain's safe rescue.

SEALS train constantly for boarding vessels or installations such as oil platforms.

US NAVY SEALS

Pakistan
2011

MAP OF PAKISTAN

After the terrorist attacks of 9/11, their organizer, Osama bin Laden, leader of al-Qaeda, became America's number-one enemy. In spring 2011, reports suggested he was hiding in Abbottabad, Pakistan. On May 2, the US government sent in Red Squadron of the Naval Special Warfare Development Group, better known as SEAL Team Six.

A CROWD mills around outside Bin Laden's compound on the day after the SEAL raid.

NIGHT RAID

The plan was to use two Black Hawk helicopters to land at night inside Bin Laden's compound. The mission was rehearsed again and again in the United States, but it went wrong in the first minute when a Black Hawk crashed. As they are trained to do, the SEALs adapted the plan. In an 18-minute operation they fought their way into the fortified house, killed Bin Laden, and left with his body before the Pakistani authorities even knew they had been there.

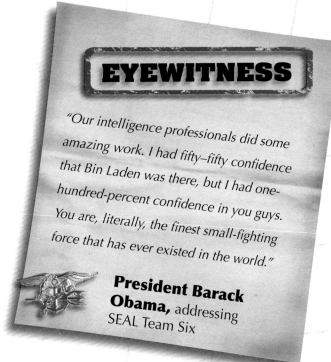

EYEWITNESS

"Our intelligence professionals did some amazing work. I had fifty–fifty confidence that Bin Laden was there, but I had one-hundred-percent confidence in you guys. You are, literally, the finest small-fighting force that has ever existed in the world."

President Barack Obama, addressing SEAL Team Six

PRESIDENT OBAMA, officials, and senior commanders watch the mission from the White House.

US NAVY SEALS

A SEAL TEAM powers through the water in a rigid-hull inflatable boat.

Boats

The SEALs have used the Zodiac Combat Rubber Raiding Craft (CRRC), or "rubber duck," since World War II. It's simple and cheap, but it gets the SEALs silently to where they need to be. The Zodiac can even be launched from a low-flying airplane.

EYEWITNESS

"You get in there, you've got all this weight on you, you close the bottom hatch, you stand doubled over—tank on your back, equipment in your arms—three to five people jammed in there. Then somebody has to find a way to reach over to the valve so you can start to flood the chamber. I've seen guys panic in there."

Bob Gormly
SEAL captain, on SDVs

BOATS

MEMBERS OF SDV Team-2 prepare to launch a minisub from the back of a submarine.

RANGE OF CRAFT

SEALs use all kinds of vessels for getting through the water, from the submersible SEAL Delivery Vehicle (SDV) to high-speed powerboats like the MK V Special Operation Craft (SOC). The newest SEAL boat is the Special Operations Craft-Riverine (SOC-R), which is designed for use on rivers and can be delivered by airplane or helicopter. The Rigid-hull Inflatable Boat (RHIB) is used to move SEALs between offshore boats and beaches.

MK V SPECIAL Operations Craft speed across the sea during a training exercise.

US NAVY SEALS

Aircraft

Aircraft transport SEALs to remote and hard-to-reach places. Sometimes, the SEALs jump from an airplane or helicopter into the ocean or a river with their rubber boats. At other times, aircraft take SEALs as close as possible to a land target so they can parachute or "fast rope" in.

BLACK HAWK

In the mission to kill Osama bin Laden in Pakistan in 2011, a SEAL Black Hawk was damaged as it landed. The SEALs destroyed it as they left. They did not want the enemy to learn their secret adaptations to the aircraft.

AIRCRAFT

ROTORCRAFT IN ACTION

The SEALs adapt Black Hawk and Pave Low helicopters to get into enemy territory. The choppers are fitted with silencers and fly without lights. Chinooks can transport men or equipment, but their size makes them a potential target. They are used sparingly in hostile territory like Afghanistan. SEALs "fast-rope" or rappel to the ground from 50 to 90 feet (15–38 m) up. To brake, the SEAL uses his hands; using his feet would damage the rope.

SEALS "fast-rope" from an MH-53 Pave Low helicopter onto the deck of a US warship.

A SEAL TEAM rushes from an Army CH-47D Chinook in a "two-wheel landing" exercise.

Specialist Gear

SEAL missions depend on surprise. SEAL teams are usually far fewer in number than the enemy, so they avoid confrontation if they can. They travel as silently as possible. Sometimes they use vehicles like jeeps. Often they swim ashore or parachute in and simply walk to their destination.

SEALS TAKE cover behind a Humvee armed with a Browning Mk 2 machine gun.

SPECIALIST GEAR

UNDERWATER

To stay underwater for long periods, SEALs use closed-circuit systems that recycle oxygen. In deeper, colder water, they use a dry suit and a larger oxygen tank. For even longer missions, the tank will carry a mixture of oxygen and air that lets the SEAL stay deeper for longer.

MINIMUM GEAR

SEALs don't wear bulletproof vests, which can restrict movement, but their helmets have headsets, and goggles for nighttime vision. On their backs are hydration packs, so they can drink without using their hands. A small first-aid kit is vital to treat wounds immediately, as is a Global Positioning System (GPS) to track location. But SEALs are trained to rely on each other rather than special gear, so they can travel light.

A SEAL drives a Desert Patrol Vehicle (DPV) in Kuwait in 2002.

43

US NAVY SEALS

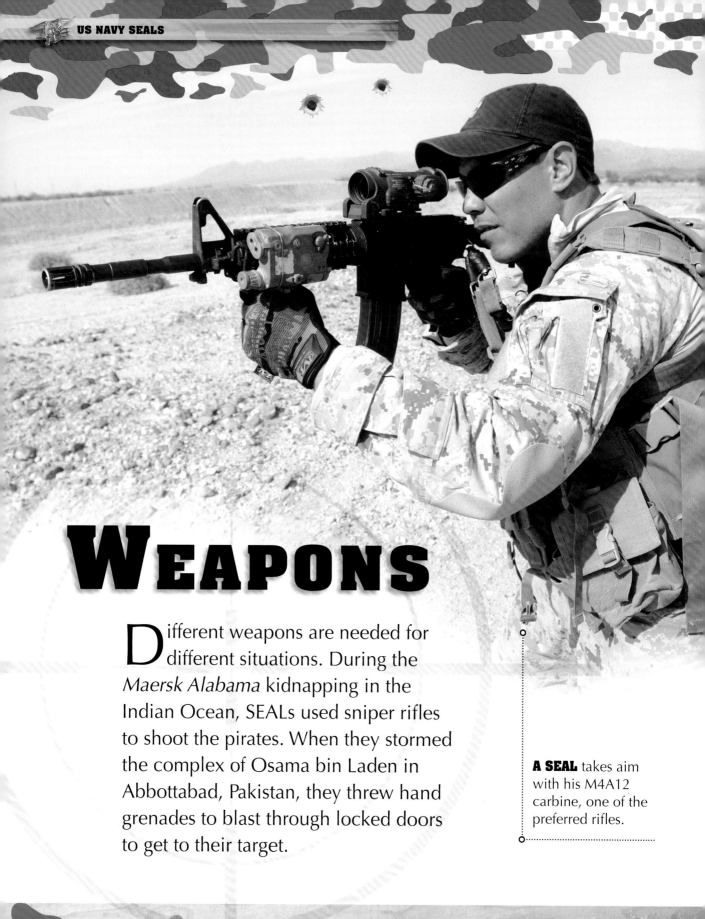

WEAPONS

Different weapons are needed for different situations. During the *Maersk Alabama* kidnapping in the Indian Ocean, SEALs used sniper rifles to shoot the pirates. When they stormed the complex of Osama bin Laden in Abbottabad, Pakistan, they threw hand grenades to blast through locked doors to get to their target.

A SEAL takes aim with his M4A12 carbine, one of the preferred rifles.

WEAPONS

THE SEALS' favored handguns include the Ruger MkII .22 pistol.

GUNS AND EXPLOSIVES

As well as sniper rifles, such as the M88-50PIP and M14s, SEALs use rifles like the automatic M4A1 carbine or even the Russian AK-47, fitted with silencers. They also use mortars, hand grenades, and antitank rockets. Sometimes, a knife does the job. SEALs are highly skilled combat fighters and can kill with their bare hands if necessary.

EYEWITNESS

"We need a variety of weapons on the teams. Certain operations call for specific capabilities. I don't think every guy should have a different weapon, just because of his own personal preference. I should be able to switch magazines with the guy next to me if I run out and he hasn't."

SEAL Team commander

HEAVILY camouflaged, a SEAL sniper waits patiently for a target.

GLOSSARY

ambush (AM-bush) To attack the enemy by setting a surprise trap.

arms dump (ARMZ DUMP) A place where large quantities of weapons and ammunition are stored.

combat swimmer (KOM-bat SWIM-mer) A soldier or sailor trained to operate on or near the coast.

commandos (kuh-MAN-dohz) A military unit trained for hit-and-run missions.

drone (DROHN) An unmanned military aircraft used for reconnaissance and for attacks on the enemy.

free fall (FREE FOL) A parachute jump in which opening the parachute is delayed until the latest possible moment.

frogman (FRAWG-man) A person equipped to stay underwater for long periods.

hostage (HOS-tij) A person who is held captive to raise money through a ransom.

inflatable (in-FLAY-tuh-bul) A type of vessel where the rubber hull is filled with air.

insertion (in-SER-shun) The act of getting forces where they need to be for a mission.

maritime (MAR-ih-tym) Something related to the sea.

mines (MYNZ) Floating bombs that are detonated if a ship passes close by.

rappel (ruh-PEHL) To descend quickly by sliding down a rope, using friction to stop.

reconnaissance (rih-KAH-nih-zents) To learn as much as possible about an enemy's positions and strength.

scuba (SKOO-buh) An abbreviation standing for Self-Contained Underwater Breathing Apparatus.

submarine (SUB-muh-reen) An enclosed vessel that can operate on top of or underneath the water.

submersible (sub-MER-sih-bul) A small craft that can operate underwater.

torpedo (tor-PEE-doh) An underwater missile with its own motor, used to damage ships beneath the waterline.

FURTHER READING

Besel, Jennifer M. *Navy SEALs*. First Facts. Mankato, MN: Capstone Press, 2011

Jackson, Kay. *Navy Ships in Action*. Amazing Military Vehicles. New York: PowerKids Press, 2009.

Payment, Simone. *Navy SEALs*. Inside Special Operations. New York: Rosen Central, 2009.

Yomtov, Nel. *Navy SEALs in Action*. Special Ops. New York: Bearport Publishing, 2008.

WEBSITES

Due to the changing nature of Internet links. PowerKids Press has developed an online list of websites related to the subject of this book. This site is updated regularly. Please use this link to access the list:
www.powerkidslinks.com/usf/navy/

INDEX

Abottabad, Pakistan 36–37
Afghanistan 28–31
aircraft 40–41
al Qaeda 4, 30, 36
amphibious operations 5–6

Basic Underwater Demoiltion School 15–17
Black Hawks 37, 40–41
boarding techniques 9, 35
boats 12–13, 25, 32–33, 38–39
Brown Water Navy 13, 25

CH-47D Chinook 40–41
combat demolition units 5, 7
combat swimmers 4, 6, 27

Desert Patrol Vehicle 43

equipment 38–45
extreme weather training 23

fast roping 40–41

Hell Week 17
history 6–7, 24–37

insertion 22
Iraq 32–33

Kennedy, John F. 5

log test 16
Luttrell, Marcus 30–31

Maersk Alabama 34–35
Marcinko, Richard 20–21
MH-53 Pave Low 41

Obama, Barack 37
Operation Just Cause 26–27
Operation Red Wing 30–31
Operation Torch 7
organization 8–9
origins 5–7

Pakistan 4, 36–37
Panama 26–27
Persian Gulf 32

Phillips, Richard 34–35
physical screening test (PST) 14–15
piracy 34–35

rifles 44–45

SEAL prep course 15
selection 14–15
Somalia 34–35
Special Boat Teams 12–13, 18
submarines 10–11
surf passage 17

Taliban 28–31
training 18–21

underwater training 18–19

vehicles 42–43
Vietnam War 13, 24–25

War on Terror 28
weapons 44–45
World War II 5–7

48